Ambush and Counter Ambush

ISBN 0-87364-098-5
Printed in the United States of America

Published by Paladin Press, a division of
Paladin Enterprises, Inc., P.O. Box 1307,
Boulder, Colorado 80306, USA.
(303) 443-7250

Direct inquiries and/or orders to the above address.

AMBUSH

AND

COUNTER AMBUSH

This volume is adapted from the Australian Military Forces publication by the same title, originally published in 1965.

Published in the U.S.A. by Paladin Press, P.O. Box 1307, Boulder, Colorado, 80306.

PREFACE

1. The terrain of South East Asia with its many defiles facilitates the technique of ambushing. Ambushing, which is perhaps the oldest tactic in the long history of war, has in recent times become the most widely used insurgent technique in Communist revoluntionary warfare. This is because successful ambushes achieve results far out of proportion to the forces deployed and permit the ambushers to melt away quickly before normal retaliatory action can be mounted effectively.

2. Thus, besides studying the enemy's ambush tactics which are described in *"The Enemy 1964"*, all Australian soldiers must master counter ambush techniques so that the tables can be turned speedily on the ambusher. It is even more important that we should acquire the ability to set and execute a successful ambush, and so exploit to the full the principle of surprise and the advantages of the terrain which ambushes afford in both limited and cold war situations.

3. This pamphlet has been compiled to provide a comprehensive guide to officers and non-commissioned officers of the Australian Military Forces and the Cadet Corps on the very important subjects of ambush and counter ambush. It contains little that is new because it has drawn heavily on the lessons hard-learnt by practical experience in areas where all Australian soldiers must be prepared to fight.

4. Of necessity, other arms and services will rarely attain the standard required of the infantryman, but this should not prevent all from striving for this standard provided they can recognize it. *"Ambush and Counter Ambush"*, which provides a common doctrine for all arms and services, enables this standard to be recognized. The doctrine contained herein supplements both *"The Division in Battle"* series of pamphlets and the pamphlet *"Patrolling and Tracking"*.

CONTENTS

CHAPTER 2 — COUNTER AMBUSH

CHAPTER 1

AMBUSHES

SECTION 1 — GENERAL

Introduction

1. An ambush is a surprise attack by a force lying in wait upon a moving or temporarily halted enemy. It is usually a brief encounter and does not require the capture and holding of ground.

2. The ambush is undoubtedly one of the oldest stratagems of battle. Even the invention and refinement of modern weapons, motor vehicles and aircraft have had little effect on the ambush pattern and techniques. The attack from ambush in ancient days was an action at close quarters, and it still is, despite the range of modern weapons.

3. Ambushes may be used in front of, and behind the enemy FEBA, against both regular and insurgent forces. A series of successful ambushes will make the enemy apprehensive and cautious in movement. Continued success will finally inflict a virtual paralysis on the enemy.

Ambush Sites

4. Suitable places for ambush include:

a. Known enemy routes in forward and rear areas.

b. Administrative areas, supply and water points. In counter insurgency operations known or suspected food dumps and arms caches are particularly suitable.

c. Areas where a marked change of vegetation occurs, such as the junction of rain forest and grassland.

d. Probable lines of enemy withdrawal after a successful attack by our forces.

Types

5. There are two types of ambush:

a. Deliberate.

b. Immediate.

6. *The Deliberate Ambush.* A deliberate ambush is one planned and executed as a separate operation. Generally time will allow planning, preparation, and rehearsal in great detail. The deliberate ambush may vary in size from a small four man ambush to a major operation using an infantry battalion. Many opportunities will exist for small scale ambushes; the ambush of up to platoon strength is the normal size used. Successful large scale ambushes will be more difficult to achieve but every opportunity must be taken. Some examples are:

a. In counter insurgency operations luring an enemy follow up force into a prepared ambush position.

b. When information provides exact enemy locations, such as an assembly area, or the movement of large numbers of reinforcements.

The smaller the force the easier it will be to introduce it into the ambush area, to control the operation and to extricate the ambush force after contact.

7. *The Immediate Ambush.* An immediate ambush is one set with a minimum of planning to anticipate imminent enemy action, or as a purely defensive technique by a force such as a patrol. In these circumstances little or no time will be available for reconnaissance before occupation. The degree of success achieved will depend on the initiative of the commander concerned, prior rehearsals and team work. A suggested drill for an immediate ambush by a patrol is described in Annex A. It should normally only be laid in areas where civilian movement by night is prohibited.

Principles

8. Instantaneous co-ordinated action against a surprised enemy held within a well covered killing ground is essential for success. This requires:

a. Sound intelligence.

b. Careful planning, reconnaissance and rehearsal.

c. Security in planning, preparation and execution.

d. Concealment.

e. Good control.

f. A simple clear cut plan for springing the ambush.

g. Maximum use of fire power. This includes accurate shooting.

h. A high standard of battle discipline, particularly by night.

9. Very careful preliminary training is required as only well trained troops with the highest standards of camouflage, concealment and fire discipline can hope to achieve success. This must be impressed on all ranks engaged in ambushing. Once in their fire positions, soldiers must remain motionless with safety catches off, and refrain from scratching, slapping at insects, smoking, eating, drinking or easing themselves. An ambush can be ruined by the accidental discharge of a weapon or by an excited soldier firing before the order is given. Unless this standard of battle discipline can be reached in all aspects, it is useless to embark on ambushing at all.

10. Sections 2 to 4 deal with points of planning and execution common to ambushes of all sizes. Points particularly relevant to small and large scale ambushes are covered in Sections 5 and 6.

11 to 12. Reserved.

SECTION 2 — PLANNING

General

13. The ambush commander will normally be told the aim of the ambush. He may be told or decide himself the strength, general location and duration of the ambush.

14. The strength of an ambush must be kept to the minimum required to achieve the aim. It may be necessary to provide reliefs if the ambush is likely to be prolonged. This will affect the overall size of the ambush force but not the strength of the ambush party. An increase in automatic weapons might allow a reduction in strength of the ambush party.

15. The extent to which the ambush commander can complete his planning before leaving his base will depend upon the availability of information and his knowledge of the ground. His planning will not be complete until he has conducted a detailed reconnaissance.

Factors

16. The following factors must be considered when planning an ambush:

a. The nature of the task.

b. The enemy.

c. Friendly forces.

d. Surprise.

e. Fire support.

f. Ground.

g. Obstacles.

h. Control.

i. Security.

j. Equipment.

k. Grouping.

l. Communications.

17. *The Nature of the Task.* Different tasks will require changes in ambush technique. Examples of tasks are:

a. Annihilation of a large force.

b. Capture of prisoners.

c. Destruction of small parties.

d. Destruction of a particular part of a column.

18. *The Enemy.* A detailed knowledge of enemy organization and likely tactics will greatly influence the layout and conduct of the ambush. Of particular interest are:

a. The likely enemy method of movement.

b. Known enemy counter ambush techniques.

c. Size of work parties, ration parties and similar detachments.

d. The pattern of enemy defensive and harassing fire tasks.

e. System of escort, eg, when escorting senior officers or supplies.

19. *Friendly Forces.* The safety of other troops moving in or near the area must be considered. Assistance, if required, may be available from either an established position or patrol base.

20. *Surprise.* The success of an ambush is dependent upon complete surprise being achieved. All planning must aim at achieving surprise. Means include:

a. Selection of unlikely ambush sites.

b. Avoidance of a set pattern in layout and techniques used in the ambush.

c. Attention to security.

21. *Fire Support.* In the larger scale ambush, guns within range and air support should be employed where possible. They are

particularly suitable for delaying interference by enemy relief forces, and to supplement cut off forces.

22. *Ground*. It may be possible to select a site by careful study of air photographs and recent patrol reports. However, a detailed reconnaissance of the following is essential to select:

a. Covered lines of approach and withdrawal.

b. Cover from view within the ambush area.

c. Suitable locations for OPs.

d. Likely enemy escape routes.

e. Suitable fields of fire.

f. Acceptable base area in case of a prolonged ambush.

23. *Obstacles*. Maximum use should be made of obstacles both natural and artificial. Obstacles may consist of a series of anti-personnel mines, including claymores, sharpened stakes, deep ditches, barbed wire or any device which will either delay or inflict casualties on the enemy. Possible places for obstacles are:

a. On likely enemy lines of withdrawal.

b. In dead ground which is difficult to cover by fire from the ambush site.

c. In areas where the main body of the enemy is likely to halt.

24. *Security*. The enemy will have a good intelligence system. The intention of the ambush force must, therefore, be disguised from the start either by moving out to the position by dark or by a well thought out cover plan. The telephone should never be used to discuss ambush plans. The need for secrecy affects prior reconnaissance of the ambush site. The operation will be prejudiced if our force is seen in the area by either enemy or civilians. In counter insurgency operations the only safe course is to assume that all civilians are potentially hostile.

25. *Control*. Good control is essential, and is difficult to achieve, particularly in large scale ambushes. The following must be planned and known to all members of the ambush party:

a. Detailed deployment into the ambush position.

b. The signal for opening fire. This will normally be controlled by the ambush commander who should be located in a suitable position to decide when the bulk of the enemy main body is within the killing ground. A deputy commander should always be nominated.

c. The signals for ceasing fire and for withdrawing to the RV.

d. Alternative arrangements to be used if the ambush is detected or otherwise compromised.

26. *Equipment*. The equipment required depends on the task and duration of the ambush. Special items which might be needed include:

a. Mines, booby traps, wire and explosives to block or canalize enemy movement.

b. Flame throwers.

c. Defence stores.

d. Flares.

e. Nylon rope, and gags to help in the capture of a prisoner.

f. Aids to tree climbing such as climbing irons for observers and snipers.

27. *Grouping*. An ambush is made up of a number of groups. The size of such groups will vary but each group should be self contained and a leader nominated. (See Sections 5 and 6). Arrangements must be made for rest as it is not possible for men to remain alert in a fire position for lengthy periods. Some men in a group will be nominated to listen and watch, while the others rest in the ambush position. Rest means that a man relaxes in his position, resting his eyes and ears but does not fidget or doze. The deployment of groups in the ambush area will be confirmed only when the final plan is made. However, during planning a broad allocation of groups can usually

be assessed. A reserve must be detailed and should normally be sited on the same side of the ambush position as the rendezvous to be used for withdrawal.

28. *Communications.* These will vary according to the size of the ambush. In a small ambush hand signals will suffice whereas in a large ambush, line, radio and light signals might all be employed. As far as possible, signal systems should always be duplicated. Means of alerting individual members whilst in the ambush are also required. A length of vine or cord tied between members has proven successful.

Layout

29. There are two fundamental principles:

a. *All Possible Approaches Must be Covered.* Information may frequently give the destination of the enemy but will rarely give the exact route they will take. No matter how good the information the enemy may arrive from an unexpected direction. This factor causes a high failure rate in ambushes. It is essential that all possible approaches be covered.

b. *The Ambush Must Have Depth.* At the first burst of fire, the enemy scatter rapidly and the chances of getting a second burst from the same position are small. Therefore, withdrawal routes must be covered to provide an opportunity for subsequent fire at the enemy.

30. Setting an ambush on both sides of the killing ground has the advantage of preventing enemy escape. It may also be necessary with a large ambush to prevent it becoming too extended. However it has the following disadvantages:

a. The killing ground must be crossed.

b. Danger to own troops.

c. Difficulty of control, particularly if a change of plan is required.

31. If an ambush is set on one side of the approach only, control is easier. However, some of the enemy will attempt to

escape from the opposite side. To prevent this stops may be placed in dead ground, or anti-personnel mines, booby traps and sharpened bamboo stakes laid across likely escape routes.

32. It may be possible to achieve the aim of an ambush by using a minimum number of men and covering the selected killing ground with anti-personnel mines such as the claymore and other controlled explosive devices. Once the weapons have been emplaced the ambush can be sprung by one or two men. This layout is particularly effective against targets about which accurate information is known.

Action After Springing the Ambush

33. This action is governed by:

a. The nature of the task.

b. The anticipated enemy strength, his deployment and likely reaction.

c. The ground.

d. Standard of training of the troops taking part.

34. The force may complete its task without moving forward from the ambush site until withdrawal. This stationary technique is suitable when long fields of fire are available or there is little threat of envelopment. It can be used in close country, but for security it requires added depth and flank protection. It must be used when visibility or the going prevent manoeuvre after springing the ambush.

35. In some circumstances it may be necessary to assault immediately after springing the ambush. The following factors should be considered:

a. An enemy will always be alert for ambush. He could be warned by unusual or suspicious movement, sounds or smells. Once an ambush is sprung, the enemy may react as follows:

(1) The leading elements, at least, dive for cover.

(2) The leading elements, if following a practiced contact drill, assault into the ambush.

Either reaction should occur in a matter of seconds. In the first case, the enemy will initially be confused due to surprise, casualties, noise and possible inability to locate the source of fire. There will be some loss of control. Panic may result, but with a well trained enemy, counter ambush drills or preparation for a more deliberate assault may be expected after the initial confusion.

b. The immediate assault must commence immediately after the initial firing which should be restricted to seconds. This speed of action is essential so that the initial period of shock created in the enemy is further intensified by the assault.

c. Even if the enemy is about to initiate, or has begun, his automatic counter ambush drill, it will usually be frustrated by an immediate assault.

36. The plan for an immediate assault from ambush may include passing directly through the enemy to an RV on the far side. Limitations in employment of this technique are:

a. The ambush layout must be single sided and as close as possible to a straight line.

b. The ground on the far side of the killing ground, as well as the approach to it, must be free of obstacles, both natural and artificial.

c. This technique is unsuitable if the aim of the ambush is the deliberate capture of a prisoner.

37. An immediate assault not involving passing directly through the enemy is only suitable when the enemy force is small. It has the fundamental advantages of exploiting the surprise created, saving own casualties and providing an opportunity to search enemy dead. The problem of control is more difficult. This type of assault is aimed at complete destruction of a small force by closing with the enemy and mopping up thoroughly.

38. The deliberate capture of a prisoner by an ambush requires a detailed plan. The action after springing should provide for the following:

a. Isolating the prisoner by fire and by movement if necessary.

b. Seizing and securing the prisoner.

c. A sound plan for evacuation of the prisoner.

39. The members of each group must be nominated and groups practised in the technique.

Withdrawal

40. The route and method of withdrawal will have an obvious bearing on the selection of the ambush site and frequently on the detailed layout. Detailed plans for the withdrawal must be made. This may involve specifying the sequence of movement of groups. When an ambush is sprung and it is quite obvious that the aim of the ambush cannot be fully achieved, the ambush commander must make full use of the surprise and temporary confusion achieved to inflict the maximum amount of damage on the enemy, and at the same time ensure a clean break. Under some circumstances small parties may be left behind to cover the withdrawal and to ambush any enemy relief force moving into the area. Plans for withdrawal must cover the following circumstances:

a. After springing.

b. Where no enemy enter the ambush site and the ambush is not sprung.

c. Where too large an enemy force approaches or enters the ambush site.

Administration

41. A large number of ambushes are sprung within a few hours of setting and require no administration other than arrangements for rest within groups. These are called short term ambushes and are the normal ambush. Where ambushes are set for periods of more than twelve hours they become long term ambushes

and administration arrangements for relief of groups for feeding and sleeping are necessary. Such an ambush may be placed on the approaches to an insurgent cultivation area which is ready for harvesting, or on the approaches to a known enemy water point.

42. In long term ambushes an administrative area must be set up. It should be sited at least 500 metres from the ambush position, far enough to avoid noises and smells disclosing the presence of troops. Water should be available. Routes to the ambush site may have to be cleared and swept to enable silent reliefs to be carried out. Although the whole party in the ambush will eventually be relieved, only one fire position should be changed at a time in case enemy approach during this period. The reliefs should take place when no enemy movement is expected. The ideal is that ambushes should be divided into three parties, one in the ambush position, the reserve, and the party at rest. On relief the party of rest takes over the ambush position; the men in the position go to the reserve; and reserve goes to the rest area.

43. The ambush commander must also consider his plan for casualty evacuation. This will depend on:

a. The remaining active strength of the force.

b. The nature of the casualty.

c. The distance to the nearest friendly location.

d. Available areas for possible helicopter evacuation.

Alternative Plans

44. Alternative plans should be avoided if possible unless the plan to be adopted in particular circumstances is simple and obvious to all members. The ambush commander must devise a fool-proof method of informing everybody of a change of plan. This will always be difficult.

45. If the enemy surprises the ambush by appearing from an unexpected direction or in unexpected strength the ambush commander must decide either to withdraw secretly or open fire and rely on surprise to make good his withdrawal. In either case he may consider the setting of another ambush along the withdrawal route to delay and inflict more casualties on the enemy.

46 to 49. Reserved.

SECTION 3 — PREPARATION AND OCCUPATION

50. *Sequence.*

a. Reconnaissance.

b. Issue of preliminary orders.

c. Preparation and rehearsal.

d. Move to the ambush area.

e. Final reconnaissance.

f. Final orders.

g. Occupation.

Reconnaissance

51. If possible the ambush commander should carry out a reconnaissance of the ambush site prior to the issue of preliminary orders. This will often be impossible and initial reconnaissance will be confined to a study of air photographs, maps and patrol reports.

52. During his reconnaissance the ambush commander should not walk on the killing area, as foot marks or disturbed earth may warn an alert enemy. Therefore reconnaissance must be done from the rear of the selected ambush site. Observing the ground from the enemy point of view, though desirable, may prejudice security. The commander will select or confirm the following:

a. *Killing Ground and Ambush Position.* A killing ground of 60 to 100 metres is desirable.

b. *Position of Each Group to be Deployed.* Locations must offer:

(1) Concealment.

(2) A good view of the killing ground.

(3) All round defence.

c. RV and routes to and from it.

d. Administration area, if necessary.

e. Withdrawal route.

53. The site selected should:

a. Be easy to conceal, so that from the enemy point of view it appears unoccupied.

b. Not offer an early escape to those enemy not killed when the ambush is first sprung.

c. Allow sentries to give due warning before the first enemy enter the ambush.

d. Be capable of being covered by all weapons.

e. Have a good covered approach avoiding contact with known enemy positions or local inhabitants.

54. *Detailed Siting.* After deciding on the general layout the commander must now consider the following points in detail:

a. *Positions of Automatic Weapons.* These must cover the killing area with subsidiary roles of sealing each end of the ambush and covering likely enemy withdrawal routes.

b. Ground not covered by automatic weapons must be covered by riflemen.

c. If enemy vehicles or tanks are expected:

(1) What blocks are required, for example trees and banks to be blown down.

(2) The position of mines.

(3) The position of anti-tank weapons.

d. Careful selection of sentry positions covering enemy approaches, to alert the ambush before the enemy reaches the killing area.

e. Likely enemy escape routes should be covered by cut off groups. Artillery, mortars, electrically-detonated grenades and mines may also be used for this task.

f. Grenades and mines may also be sited to protect the flanks and rear of the ambush against quick enemy counter action. For the protection of the ambush party they should be sited in defiladed ground.

Orders, Preparation and Rehearsals

55. *Preliminary Orders.* The ambush commander should brief his party with the aid of a model as thoroughly as possible to reduce the time spent on final orders, and as early as possible to allow the maximum time for preparation and rehearsal. An orders aid memoire is attached as Annex B.

56. *Preparation.* Thorough preparation is essential. Extra automatic weapons will normally be needed and because of the great reliance on heavy and immediate fire, all weapons must be thoroughly cleaned, checked and tested to ensure their efficiency. Special stores, such as trip flares, may have to be assembled and tested.

57. *Rehearsals.* The rehearsal must:

a. Show troops their positions relative to each other.

b. Show how reliefs, if any, will take place.

c. Cover the springing, assault and withdrawal phases.

d. Eliminate any tendency to fire high or failure to aim when firing.

e. Develop team work.

f. Test communications and signals.

The rehearsals should aim at saving time and lengthy orders when the actual ambush site is reached. Final rehearsals for night ambushes are conducted at night. If illumination devices are to be used in the ambush they should be used in the rehearsal.

58. *Move to the Ambush Area.* The main body should not move direct to the ambush position. It should assemble short of the ambush position, possibly at the RV, whilst the reconnaissance party goes forward to see that no enemy are present and, if it has not previously been done, to carry out a reconnaissance.

59. *Final Orders.* If reconnaissance is carried out before preliminary orders are issued there should be only need for brief confirmatory orders unless last minute changes are necessary. If preliminary orders are issued before reconnaissance, the

outline plan needs only modification from any additional information obtained by reconnaissance. In any case final orders must be brief and include:

a. A description of the ambush area and killing ground.

b. Final location of the commander.

c. Any variations from the rehearsal in regard to individual tasks.

In the event of a night arrival at the ambush site, the immediate reconnaissance must be confirmed at first light.

Occupation

60. Individual camouflage must be checked before moving forward to occupy the position. Since the enemy may move into the killing ground as the ambush is being laid, occupation must be carried out stealthily from the rear with only a few men moving at a time.

61. The normal sequence for occupation is:

a. Sentries or observers take up their positions and communications are established with the commander.

b. Automatic or anti-tank weapons are brought forward to cover the killing area.

c. If used, flares, mines, grenades and charges are set.

d. The remainder of the party, including troops in depth for rear and flank protection, and cut off parties are placed in position.

e. Reliefs, if any, are shown the ambush site and then moved back to the base or administration area.

62. Care must be taken to avoid giving away the ambush to the enemy. Particular attention should be paid to:

a. Paper scraps, foot prints, bruised vegetation, trip wires, reflecting surfaces.

b. Items with a distinctive smell should be left behind. Men's hair should be washed free of hair oil and hair creams; cigarettes should be withdrawn; sweets, chewing gum and other food, including curry powder, must not be carried.

c. Weapons, which must be cocked and in a state of instant readiness to fire during the wait.

d. Any civilians who are suspected of having discovered the ambush and who should be held where they can not give warning until after its successful completion.

63. Each individual soldier in the ambush area must be responsible for:

a. Personal camouflage.

b. Taking up the best available firing position.

c. Remaining still and silent for protracted periods.

64. In a small scale ambush the force can normally be deployed immediately into the ambush site. However, where the enemy are known to deploy leading elements to search ground in likely ambush areas, the ambush force must lie back from the chosen area in hides and only move forward on signals from OPs. This is normal with large scale ambushes. (See Section 6).

Lying in Ambush

65. Troops must be trained to select a comfortable position and remain in it without smoking, undue movement or noise, for the whole of the time they are in the actual ambush site. This may be some hours. Specific orders must be given concerning eating and drinking.

66. Weapons must be cocked before moving into position, and safety catches left off.

67. If all members of the ambush observe continuously, no rest is possible and keenness will deteriorate. On the other hand, the killing area must be under observation at all times. This is achieved by rostering observers within groups, with group leaders nominated as required. The ambush commander and his second in command relieve each other. Other members relax but remain so that, without undue movement, they can fire on their arcs as soon as they are alerted.

68. Whether the ambush party will need relief depends on the number of troops available and the duration of the ambush. Reliefs are made only when essential, but troops should not be left in an ambush site too long merely to avoid the problems of relief. In still air conditions, when the temperature and humidity are high and there is no effective shade from the sun, the alertness and efficiency of troops will deteriorate rapidly, possibly to the extent that security is threatened, unless counter measures are taken. These conditions are typical of crops, grasslands and areas of low scrub on clear bright sunny days in the humid tropics. To safeguard the effectiveness of an ambush in these circumstances, the troops must be acclimatized and relieved regularly before deterioration sets in. Adequate water should be provided for the ambush period. Reliefs must be planned, each man being relieved, quietly and slowly one at a time. (See Paragraph 42).

69 to 72. Reserved.

SECTION 4 — EXECUTION

Springing

73. When a sentry sights the enemy he tugs the communication cord or gives the signal for the direction of approach and size of the enemy party. When the enemy appears in the killing ground each man will aim awaiting the order to fire. The ambush should be sprung when as many enemy as possible are in the killing ground and the range has been reduced to a minimum. There must be no half-heartedness or premature action. All men must clearly understand their orders, and the drill for opening fire, as follows:

a. Fire should not be opened so long as the enemy is moving toward someone in a better position to kill.

b. A small ambush will normally be sprung by the commander. but should any enemy act as though he has seen the ambush, any man who sees this action should spring the ambush. Because of the risk of the ambush being prematurely sprung, only well trained and experienced soldiers should be sited in those positions close to the killing ground where an alert enemy could discover the ambush.

c. All shots must be aimed to kill. Once fire has been opened men may have to stand up to fire at moving targets.

74. The commander should be so placed that he has a good view of the enemy. The signal for springing may be one of the following:

a. An aimed burst from either the commander's weapon, or from a machine gun which he controls.

b. The controlled explosion of a grenade necklace.

c. The setting off a booby trap or trip flare.

75. The commander must always make alternative arrangements for springing the ambush in the event of something going wrong with the person or method nominated. A deputy must always be appointed and the chain of command in seniority must be known.

Subsequent Action

76. Once the ambush is sprung, subsequent action proceeds according to the plan, ie, either an immediate assault or remain in the ambush positions. (See Paragraphs 33 to 36). Normal action is to fire from positions until the enemy has been destroyed. Only then would assault parties move in to clean up and search for prisoners and documents. The fire fight is normally of short duration. The signal to stop firing will be given by the commander — a whistle blast is suggested.

77. The search group normally:

a. Checks for enemy in the killing ground and secures any who are still living.

b. Searches the surrounding area for dead and wounded.

c. Collects the enemy's arms, ammunition and equipment.

d. If required, photographs the bodies for identification.

78. If the aim of the ambush is to secure a prisoner: specific members of the ambush party must be detailed in orders, rehearsed and equipped with adhesive tape gags, garottes, signal cable or toggle ropes to ensure that prisoners are seized and

escorted quickly and quietly to the RV. Reserves should also be detailed for these tasks to provide for possible casualties.

79. *Tracker Teams.* Some enemy, who have been wounded in the ambush, may attempt to escape by rushing into the undergrowth and lying low until the ambush has withdrawn. If the operational situation permits the employment of trackers will quite often lead to their capture.

Withdrawal

80. It is at this stage that the ambush is most vulnerable. If no assault is planned, withdrawal should be made during the brief period of enemy confusion and before he has re-established control. A rendezvous is necessary for the ambush party, as members may take some time to clear the site. A check point between the ambush site and the rendezvous may be required. Where an assault, search of victims and collection of prisoners is required planning must include arrangements for the withdrawal of these groups covered by the remainder of the party.

81. The method of withdrawal will be covered in the orders and must be rehearsed. Troops make a clean break from the ambush site, and concentrate at the rendezvous quickly and in an orderly manner. Time in the RV must be short; the force must be checked, formed up and moved off as quickly as possible. More than one route of withdrawal may be necessary. It may be desirable and practicable to set a further ambush along the withdrawal route.

82. A normal sequence for withdrawal is:

a. Sentries.

b. Ambush party by groups, eg, killing and search groups.

c. Troops in depth.

83. *Casualties.* Arrangements must be made before occupation for the evacuation of both our own and enemy casualties if necessary. Stretchers or material for improvising stretchers may be dumped at the RV so that the minimum time is lost during withdrawal.

Ambushes at Night

84. Night ambushes are often the most successful because enemy movement generally increases during the hours of darkness.

85. Night ambushes have similar characteristics to ambushes by day. Particular points which apply to night ambushes are:

a. Concealment is easy but shooting is much less accurate.

b. Automatic weapons become the essential fire elements, single shot weapons being slow to produce the necessary volume of fire.

c. All weapons, particularly machine guns firing down tracks, should have their left and right arcs of fire fixed by means of sticks, etc, to eliminate danger to own troops.

d. The ambush party must never move about. Any movement will be regarded as enemy.

e. Clear orders, precise fire control instructions, clear RV and signals are essential.

f. Men and groups will be sited closer together than by day. Control at night is all important.

g. It is difficult to take up an ambush position at night. Where possible it should be occupied before last light.

h. Consideration must be given to a method of illuminating the killing area at the moment the commander wishes to spring the ambush. Illumination can be provided by means of trip flares or ground marker flares, set off by the ambush commander either electrically or by pulling a trip wire. Flares may be set off by animals so it is generally better if they are detonated electrically. The flares should be sited so that they will illuminate the enemy whilst at the same time, members of the ambush are shielded from the direct glare of the light. Notes on the use of flares are included in Annex C. The use of night viewing devices such as infra red weapon sights and binoculars, or weapon torch attachments are also useful.

86 to 87. Reserved.

SECTION 5 — SMALL DELIBERATE AMBUSHES

88. This section deals in particular with the layout of small scale deliberate ambushes of up to platoon strength. The layouts described are particularly useful in counter insurgency operations.

Grouping

89. The ambush is made up of killer and cut off groups. The size of these groups will vary, but two to six men can be taken as a guide. Each group should be self contained, and a leader nominated who is responsible to the ambush commander.

Layout

90. Groups may be employed in two ways, bearing in mind the principles of layout:

a. *Area Ambush.* Where there is more than one approach all must be covered. Approaches should be covered in depth to catch enemy scattering from the position of the ambush. Such an ambush is known as an area ambush. (See Figure 1). It consists of a series of small groups, each with its own commander, sited as part of an overall plan to encompass a particular enemy party which is expected. The ambush party moves to a dispersal point from which groups move by selected routes to their positions. The ambush commander may be able to position only one group in detail, leaving the remainder to be positioned by group commanders.

Notes:

1. Information has been received that insurgents will contact farmers at or near track junction D.

2. It is decided that the enemy will probably approach through primary jungle along the side of the slope. Alternative approaches are A to D, B to D, C to D or through rubber.

3. Ambush groups are posted at A, B, C, D, E.

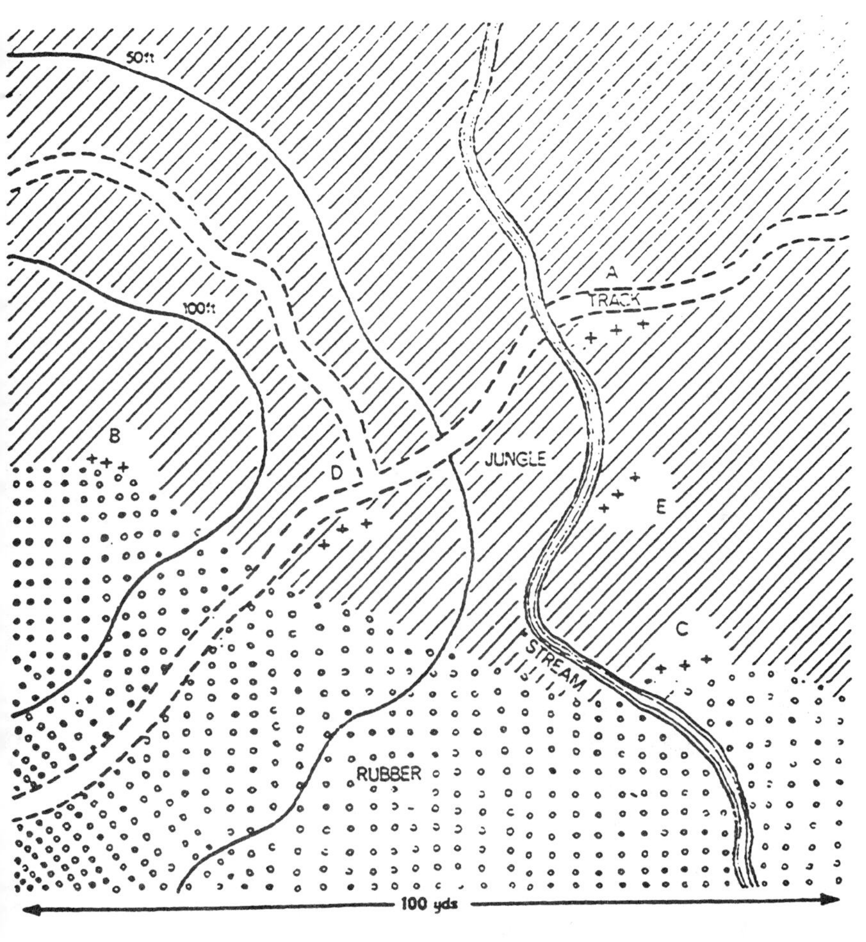

Figure 1 — Area Ambush.

4. If enemy approach from A to D, A will allow them to pass through.

5. D will probably spring ambush. Enemy will scatter and may run into A, C (both downhill) or E. If they hi' A or C they may rebound along stream on to E.

b. *Limited Ambush.* When, because of the ground, there is only one likely approach, a group or groups may be sited in depth with all round defence at a place on that route which gives adequate concealment. This is a limited ambush. (See Figure 2). It is used when the area ambush is impossible or as part of an area ambush, along a very likely approach track.

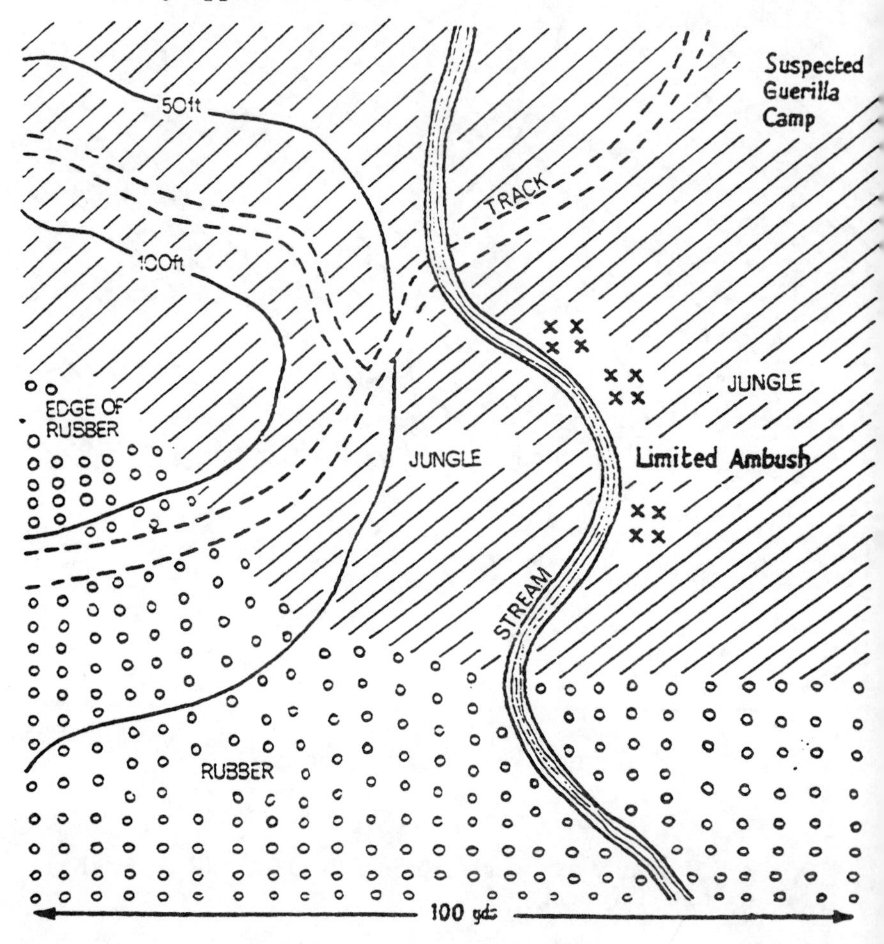

Figure 2 — Limited Ambush.

Notes:

1. A group is known to be in an area approximately 1,000 metres square.

2. Information has been received that a party of three enemy will collect subscriptions from farmers. The party will approach along the line of the stream. The ambush is therefore sited along this stream.

3. The ambush could be improved by siting the Limited Ambush as one element of an Area Ambush thus covering all approaches.

91 to 92. Reserved.

SECTION 6 — LARGE DELIBERATE AMBUSHES

Size

93. The term large scale ambush implies that the strength of the ambushing force is at least one company and possibly up to battalion. With greater numbers, greater difficulty will be experienced in achieving surprise, which is essential to success.

Problems

94. The principle problems facing the commander of a large scale ambush will be:

a. Introduction of his force into the ambush area.

b. Concealment of the elements of the ambush force.

c. Command and control, including:

(1) Timing for the springing of the ambush.

(2) Canalising the enemy lines of withdrawal.

(3) Location of his own HQ for control of all elements of his force.

(4) Co-ordination of fire between our forces.

(5) Adequate time for reconnaissance and arranging the ambush and possible rehearsal.

Reaction and Counter

95. Any enemy force that warrants the setting up of a large scale ambush will normally be moving considerably dispersed. His likely reactions on the ambush being sprung are:

a. Immediate assault to break out of the ambush; and

b. Dispersion into small groups with stay-behind parties to cover the withdrawal.

96. To counter these problems, the ambush commander must be prepared to:

a. Accept considerable dispersion between the elements of his own forces.

b. Ensure that the enemy main body is within the ambush area before opening fire.

c. Employ blocking and cut off forces of considerable size.

d. Form a mobile reserve to act offensively.

Static Ambushes

97. Where the problem of concealing the elements of a large scale ambush does not exist (as in dense jungle) and where the enemy pattern of movement permits, and his line of advance can be reasonably predicted, it is suggested that the ambush might take the form shown in Figure 3.

Notes:

1. A Coy is deployed astride the enemy's lines of advance to act as the first stopping force and prepared to fight a defensive battle in the face of enemy assault.

2. B Coy is deployed to a flank in concealed positions along the enemy line of advance to:

a. Fire into the enemy main body when the ambush is sprung.

b. Assault into the ambush if required.

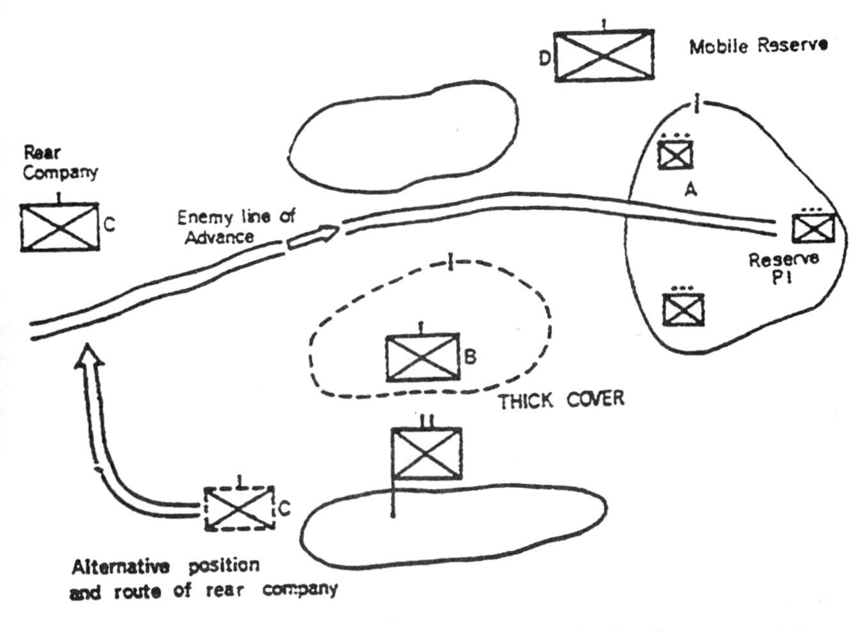

Figure 3 — Large Scale Static Ambush.

3. C Coy deployed to a flank, probably at a distance from the line of advance with the task of:

a. Acting as the cut off force for the ambush.

b. Moving to assault the enemy from the rear.

c. Preventing a relief force from reaching the ambush.

d. If full concealment is possible, this company may sometimes be in a position astride the line of advance initially but will not disclose its presence until the stopping force has opened fire.

4. D Coy deployed to the flanks at some distance as the commander's mobile reserve.

Mobile Ambushes

98. It will often happen that:

a. It is not possible to conceal an ambush of battalion strength in close proximity to the ambush site.

b. The enemy will clear his line of advance with soldiers on foot.

c. The enemy may advance well dispersed.

99. In any of the above cases, the commander of the ambush force must be prepared to lay a more mobile type of ambush.

The problem of control in such cases is greatly increased. Figure 4 shows an example of this form:

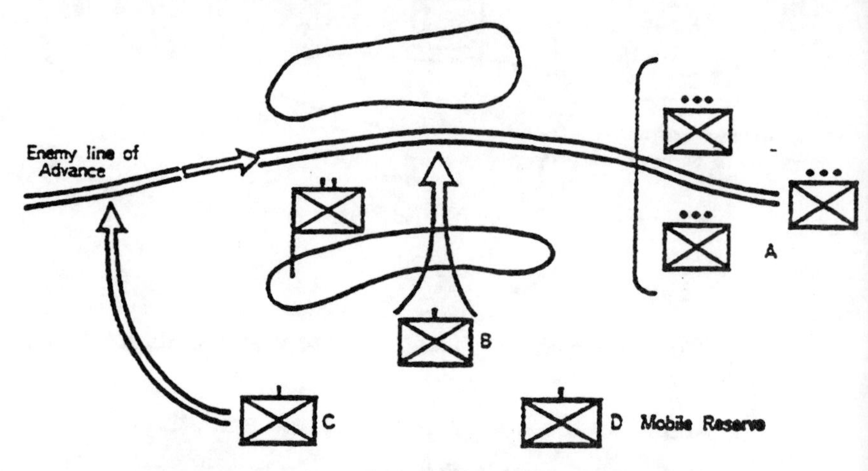

Figure 4—Large Scale Mobile Ambush

Notes:

1. The essential differences are:

a. Greater dispersion along the enemy lines of advance.

b. All companies, including the dispersed platoons of the blocking company, are deployed far enough away from the ambush for concealment and to avoid being discovered by the leading enemy elements clearing the route. They must be prepared to move in quickly to take up fire positions, or to assault into the ambush if necessary.

2. In both Figures 3 and 4, demolitions, minefields, heavy fire support (both armour and artillery) have been omitted for simplicity.

100. Springing an ambush of this type poses many problems; the solution of which will vary in each case. Some points are:

a. The commander should be concealed close to the ambush site with duplicated communications to his OPs. He may, on occasion, be located with the blocking force.

b. OPs must be carefully concealed either in trees or dug into scrub.

c. Pre-planning must be done in case of communications failing and a deputy commander appointed who must be given authority to spring the ambush in specific circumstances.

101. Reserved.

SECTION 7 — TRAINING

General

102. Ambush training must be aimed at eliminating common faults and improving techniques. Its aims are:

a. Training of troops to occupy positions without advertising their presence by footprints, movement by individuals when the enemy is approaching, and the noise of weapons being cocked or safety catches and change levers being moved.

b. Ensuring the good positioning of commanders and siting of weapons. A lack of all round observation will result in the enemy arriving in the area undetected.

c. Improvement of fire control and particularly the even distribution of fire.

d. Ensuring accurate shooting at moving targets and reduction of the tendency for men either to select and fire at the same target or high at the light faces of the enemy.

e. Improving the care of weapons and preventing misfires and stoppages occurring through failure to clean, inspect and test weapons and magazines.

Causes of Failure

103. The following are some reasons for failure which have been reported by ambush commanders and which may help in training:

a. Disclosure of the ambush by the noise made by cocking weapons and moving safety catches or change levers. Check your weapons, practise men in their silent handling and ensure that all weapons are ready to fire.

b. A tendency to shoot high at the face of the enemy. This must be corrected on the ambush range.

c. Disclosure of the ambush position by footprints made by the ambush party moving position and by the movement of individuals at the crucial time when the enemy were approaching.

d. A lack of fire control as commanders were unable to stop the firing and start the immediate follow up.

e. Commanders were badly sited with consequent lack of control.

f. A lack of all round observation resulting in enemy arriving in the area of an ambush unannounced.

g. Misfires and stoppages through failure to clean, inspect, and test weapons and magazines.

h. A lack of a clearly defined drill for opening fire.

i. A tendency for all to select and fire at the same target.

j. Fire opened prematurely.

The Ambush Range

104. The constant need for shooting practice must be emphasised. The object of having an ambush range is to practise fire control and shooting from an ambush position, in conditions representing, as nearly as possible, an operational ambush. The requirements, which are easy to fulfil are.

a. *Ambush Position.* This should be large enough for about a section and needs careful selection. Natural cover will be required and therefore the position should be left untouched as far as possible.

b. *Killing Ground.* The killing ground should look as natural as possible from the ambush position, but trenches need to be dug in order that targets and markers can be moved about. If the ground allows, there should be several trenches at different angles, so that targets may approach and withdraw from different directions. A possible layout is shown in Figure 5.

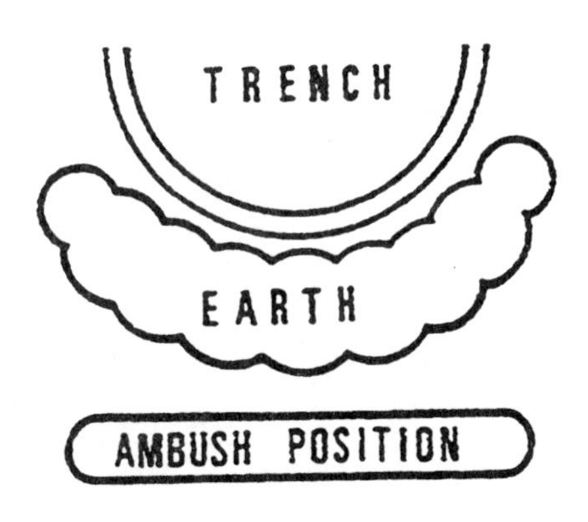

Figure 5—Layout of Ambush Range.

c. *Safety Precautions.* Care must be taken to ensure that sufficient earth is thrown up at the right places to give ample protection for the markers. If the ground does not favour natural protection, pulley raised targets should be used. Safety precautions are detailed in "*Infantry Training Volume III, Pamphlet 31, Rangework General*", and "*Pamphlet 32, Range Construction and Regulations (All Arms)*".

d. *Ingenuity*. Exercises run on the ambush range depend on realism for success. The following points are useful:

(1) All actions by troops in the ambush position must be fully operational, eg, position taken up silently, camouflage, and clear orders.

(2) A wait should be imposed to introduce realism. Targets should appear without any warning. Once fire has opened targets must move rapidly.

(3) The range should be used day and night.

105. Reserved.

CHAPTER 2

COUNTER AMBUSH

SECTION 8 — GENERAL

Introduction

1. Any military force that moves on the surface of the earth can be attacked from ambush. This includes columns and patrols moving on foot, vehicle convoys, railway trains and rivercraft. The object of the attack from ambush varies from complete destruction of the target to delaying a moving column.

2. The ambush is the most widely used insurgent technique. It enables the insurgents to overcome the over-all disparity in numbers, firepower and technology which normally exist between themselves and conventional forces. By operating under cover, by exploiting their intimate knowledge of terrain and superior foot mobility, the insurgents can choose the time and place for brief concentrations to attack small elements of our forces in transit. The capture of weapons, equipment and supplies is often a major aim.

3. Mobile reserves or relieving forces moving to counter attack or assist a defended position or post which has been attacked or raided are favourite enemy targets for ambush.

4. Insurgent forces are not alone in favouring the ambush technique. It can, and must be expected that conventional enemy forces will attempt to ambush our troops whilst moving both on foot and vehicles.

5. The characteristics and types of enemy ambush are fully described in "*The Enemy 1964.*"

Responsibilities

6. Where an insurgent threat exists, the whole operational area is vulnerable to attack and ambush. Even vehicles taking soldiers

to sporting fixtures and places of entertainment may require escorts. Movement of road convoys outside a controlled area in support of offensive operations in depth is extremely hazardous and always liable to large scale enemy ambush — particularly on the return journey. Members of all arms and services must be aware of the looming threat of ambush and must be proficient in counter ambush drills.

7. The enemy will be less confident of their ability to execute an ambush if they are unable to achieve worthwhile results from such operations, or if they suffer heavy casualties because of our counter action. The enemy ambush can be beaten by immediate positive offensive measures which are thoroughly understood and practised by all ranks of all arms and services. Counter measures can be divided into:

a. Precautions taken to avoid or detect an ambush.

b. Action on being ambushed.

8 to 9. Reserved.

SECTION 9 — COUNTER AMBUSH ACTION FOR A FORCE MOVING ON FOOT

Avoidance of Ambush

10. Any column or detachment moving outside a secure area must be prepared to counter enemy ambush tactics. The obvious measure is to avoid being ambushed. This may be done by denying the enemy foreknowledge of our movements or by detecting the ambush.

11. To avoid ambush:

a. Routine movement must be reduced to a minimum.

b. Except in completely secure areas, roads and tracks should never be used if they can be avoided.

c. Security of impending operations and movement must be maintained until the last possible moment.

d. Plan and use deception whenever possible.

e. During movement, full dispersion and movement by bounds must be practised and controlled. The degree of security of a column on the march may be regarded, roughly as inversely proportional to its rate of march.

f. Thorough reconnaissance to the front and flanks is required. Helicopters are useful for this task, landing troops to search selected areas if necessary.

g. Maps, air photographs, patrol and other reports must be studied to find likely ambush sites. These should be cleared on foot.

h. Warning of enemy ambushes may be obtained from local inhabitants. An unusual lack of civilian activity in villages, padi fields, etc, is often a sure indication of enemy activity, though the presence of "*normal*" activity should not be regarded as an indication of safety.

Breaking out of an Ambush

12. A force laying an ambush has the advantages of selection of site, initiative and surprise. The ambushed force is at a tactical disadvantage which can be minimized by good training and resolute action. The basis of the counter ambush battle is controlled offensive action.

13. It will be appreciated that there can be no hard and fast rule for breaking out of an ambush. There are, however, two essentials which are common to all counter action. These are:

a. Immediate offensive action must be taken to break out of the killing zone as rapidly as possible. At lower levels immediate action drills are used for this purpose.

b. Commanders must retain control. Alternative arrangements for command must be made in case the commander is lost in the first contact.

14. A suggested immediate action drill for use when caught by an enemy ambush is described in Annex D.

15. A different drill particularly suited to close country and restricted enemy fields of fire is an immediate assault in one

direction into the ambush. Such a drill must be planned and rehearsed prior to the action. The only orders, if any, required are *"Follow me"*, *"Charge"* or some other simple words to achieve an immediate reaction.

16. Enemy ambush tactics will vary and opportunism and initiative by group leaders and individuals will always be required in the counter ambush battle. Planning should consider the possibility of the enemy using blocking parties to counter our immediate action drill.

Action if Only Part of a Force is Ambushed

17. The portion ambushed must take aggressive action to fight its way out of the immediate killing zone using fire and movement if necessary. This part of the force then forms a hasty defensive perimeter sited if possible to bring fire on to the ambushers. Obvious reorganization positions will always be suspect as the enemy may have laid mines or booby traps in the area.

18. The remainder who are not caught in the actual ambush must do an immediate encircling attack against a flank. In so doing contact with enemy blocking parties can be anticipated. Full advantage should be taken of artillery or close air support available. If, however, this would delay the mounting of the attack, its advantages should be carefully weighed against the requirements to relieve the ambushed force.

Action if the Whole of a Force is Ambushed

19. The force must take aggressive action to fight its way out of the immediate killing zone. The following courses are available:

a. Launch an immediate assault. This is dependent upon the degree of control retained in relation to the enemy's main strength and dispositions.

b. Form a hasty defensive perimeter whilst the commander decides whether to attack part of the ambush in order to break out, or whether the force will break down into small groups to filter out. In either case, the force must reform at a pre-planned RV as soon as possible.

c. On occasions, during the immediate assault to break out of the killing zone, it may be possible to seize ground on which a reasonable defensive perimeter can be established. The aim would then be to hold this perimeter, to bring in close air support and/or artillery against the enemy position or to wait the arrival of a mobile relief force probably brought in by helicopter. Such action will often force the withdrawal of the enemy ambush and has the advantage that it prevents our own wounded and equipment falling into enemy hands.

Counter Ambush by Night

20. As night ambushes are extremely difficult to arrange and co-ordinate they are likely to be on a small scale only.

21. If our forces are ambushed:

a. They must move out of the killing zone at once. This is especially important if the area has been illuminated.

b. They must fight their way from the ambush to the pre-planned RV.

c. Commanders must retain control. If lost it must be regained as soon as possible.

d. There can be no question of a flanking or encircling attack at night because of the difficulty of control and the degree of confusion that will exist in the ambush area.

Special Points for Counter Ambush

22. The following points must be particularly remembered by commanders moving forces in areas where they are likely to be ambushed:

a. The enemy will aim in the opening volleys of an ambush to destroy commanders and radio operators in order to increase confusion. Commanders with their radio operators must not be conspicuous and must avoid moving to a set pattern within a column. The practice of commanders at all levels carrying a particular weapon, such as a sub-machine gun is dangerous. Badges of rank should not be obvious. Radio operators must be protected and unless the sets are being operated, aerials should be dismounted.

b. During movement, maximum dispersion commensurate with control, must be practised. The aim must be to ensure that the whole of the force is not simultaneously ambushed. Too often our forces are closed up, forming a *"crocodile"*, and thus making themselves vulnerable to a comparatively small ambush. This is particularly applicable to the rear of a column. The degree of dispersion practised is dependent upon the likelihood of enemy action.

c. Pre-planned RVs in the event of an ambush must be known to all ranks and, if possible, should be constant. Some suggestions for RVs are:

(1) A set distance (500 metres) from the rear of the column and back along the direction of approach. This is probably not suitable for forces larger than a platoon.

(2) The location of the last long halt.

23. Reserved.

SECTION 10 — COUNTER AMBUSH ACTION FOR VEHICLE CONVOYS

General

24. The ambushing of sub-units travelling in vehicles is not difficult and can be very costly if the troops are not alert and trained in counter vehicle ambush drills.

25. This section deals with the problems of being attacked whilst moving in vehicles, the precautions which can be taken before contact and the correct action to recover the situation whenever a sub-unit is ambushed. It should be read in conjunction with *"The Division in Battle, Pamphlet No 11, Counter Revolutionary Warfare", Sections 46 and 48* which describe the planning and execution needed to safely pass a convoy along a road exposed to insurgent action.

Action before Contact

26. *Preparation of Vehicles.* Troops travelling in vehicles must be able to see all round them, fire their weapons and throw smoke grenades over the vehicle's sides without hindrance. They

must debus quickly with the minimum of restriction of movement. For these reasons, vehicles such as 2½ ton or 5 ton trucks, should have their canopies and canopy framework removed and their tailboards down. Although vehicle canopies can be rolled up to give overhead cover only. it should be remembered that canopy framework restricts the throwing of grenades, the traversing of weapons and quick de-bussing over the sides of the vehicles. Troop carrying vehicles, and certainly the leading vehicle in each packet must be sandbagged against mines and have bolt on armour if available. Notes on the use of sandbags for vehicle protection are included in Annex E.

27. *Loading of TCVs.* Twenty men standing or sitting in the back of a truck, say 2½ tons are a very vulnerable target. In a sudden emergency they will be unable to use their arms effectively and they are liable to turn into a jumbled heap of bodies and weapons quite beyond the control of their vehicle commander. Fourteen or fifteen men is the maximum that can be carried safely in the rear of a 2½ or 5 ton vehicle.

28. *Vehicle Commanders.* A commander must be detailed by name for each vehicle. His task will be to post sentries, ensure that all personnel are alert and assist in maintaining convoy formations. He commands the troops in his vehicle should the convoy be ambushed. He must travel in the rear of the vehicle and not in the cabin with the driver.

29. *Vehicle Sentries.* Four sentries should be posted in the back of each TCV except the smaller types such as Landrovers. The two sentries at the front should cover 1600 mils from front to side and the two at the rear 1600 mils from rear to side. Where possible these sentries should be armed with automatic weapons. It is their task to take immediate action from those positions should the vehicle be ambushed and to cover the evacuation of the vehicle should it be brought to a halt. Sentries must also assist in control of the convoy by notifying the commander of any disruption in the convoy formation. In addition to the sentries posted in the body of the vehicle, an additional soldier must be detailed to sit in the front seat of the vehicle's cabin beside the driver. The task of this individual, normally called the "*Shot Gun*", is to assist the driver in an emergency in controlling or stopping the vehicle. If a vehicle is halted by enemy

action the shot gun stays with the vehicle to act as close protection for the driver, the vehicle and any stores it may be carrying. He does not accompany any assault or sweep that may be carried out by the other troops in the convoy.

30. *Heavier Weapons.* Machine guns and rocket launchers should be distributed throughout the convoy. It should be remembered that a machine gun on its bipod perched on a cabin top is in a very insecure position. A gunner holding it with two hands is very liable to be jolted off his feet should the vehicle swerve suddenly or the driver unexpectedly brake hard. The gun may well go one way, the gunner another. The rocket launcher can be used to give direct covering fire for a flanking assault on to the enemy position. The safety of our own troops must be considered before opening fire with this weapon.

31. *Armoured Fighting Vehicles.* The vehicles of a Cavalry Squadron are useful for convoy protection. They are tracked, with a good cross-country performance, armoured against small arms and fragments and are amphibious.

a. *Air Portable Tank.* This vehicle mounts a large-calibre gun and machine guns and is able to provide heavy fire against an ambush.

b. *Light Reconnaissance Vehicle.* This vehicle mounts a heavy machine gun which has an excellent neutralizing effect. Being smaller than the other vehicles it may be able to pass halted or immobilized vehicles so as to bring effective fire on to the ambushing force.

c. *Armoured Personnel Carrier.* This vehicle carries an assault section under armoured protection. The machine gun mounted on the vehicle can not be relied upon for suppressive fire in an ambush situation as the gunner is completely exposed when firing.

32. *Convoy Commander.* The convoy commander should position himself where he considers he can best control the convoy. He should nominate a vehicle commander for each vehicle and brief them thoroughly before moving from the start point. He will always nominate a successor.

33. *Briefing.* Briefing by the convoy commander before the convoy moves off must be detailed and explicit. All drivers and vehicle commanders should be present at the briefing and if possible the men travelling in the convoy should also be present.

Briefing should include:

a. Detail of timings, route, speed, density, order of march, maintenance of contact, procedure when contact is lost and action on break down.

b. Distribution of personnel to vehicles and their conduct in them.

c. Appointment of vehicle commanders, sentries and details of action on ambush for that particular convoy.

d. Action in new areas — sweep possible ambush sites on foot or use prophylatic fire from automatic weapons when approaching suspicious ground.

34. *Security.* Movement by road should never become routine. Remember that telephone and radio communications are not secure and codes, etc, should be used when discussing or detailing future road convoys by these means.

Action on Contact

35. Whatever precautions are taken and preparations made, the ambush when it is sprung, will always be an unexpected encounter. Contact drills are simple courses of action, designed to deal with the problem of the unexpected encounter. They detail immediate, positive and offensive action.

36. The enemy springs his ambush on ground that he has carefully chosen and converted into an area where he can kill his enemy by firing at them, often at point blank range from above. The principle behind the contact drill is that it is fatal to halt in the area the enemy has chosen as a killing ground and so covered by fire, unless forced to do so.

37. The drill therefore, is either to drive through the ambush area or halt before running into it, and to attack the enemy immediately from flank and rear.

Contact Drill

38. In order that the enemy may not have the advantage of operating on ground of his own choosing, every effort must be made to get vehicles clear of the area in which effective enemy fire can be brought to bear (Danger Zone). When vehicles are fired on the drivers will not stop but drive clear of the danger zone. The sentries in the vehicles will fire immediately to keep the enemy's head down. When vehicles are clear of the danger zone, they will stop to allow their occupants to dismount and carry out offensive action. Following vehicles approaching the danger zone should not run through the ambush, but halt clear of the area.

39. Where vehicles have not been able to clear the danger zone troops will debus, get off the roadway away from the enemy and make for cover. Vehicle sentries will cover the debussing with fire, including smoke if possible. The troops will then bring fire to bear on the enemy or at least cover the vehicles to stop enemy setting fire to them. They can often shoot the assaulting force on to their objective, but must be careful to stop firing when the assaulting troops reach the safety limit.

40. If the ground permits and there are no troops clear of the danger zone to outflank the enemy, it may be possible to make a frontal attack covered by smoke.

41. The enemy is sensitive to threats to his rear or flank. Offensive action to produce such a threat can only be carried out by those troops clear of the danger zone. The position of such troops can be:

a. *Not Yet Entered the Danger Zone*. The convoy commander, or, if he is not there, the senior vehicle commander present, will carry out an immediate flanking attack on the enemy position supported by fire from such weapons as machine guns and rocket launchers.

b. *Clear Ahead of the Danger Zone*. An encircling attack will be made by the troops clear ahead of the ambush. It is doubtful however, if this attack can be put in as quickly as in a. The troops are moving away from the scene of action, and may not be immediately aware of the ambush. They have to be marshalled and brought back to a starting point for this attack.

c. *Both Sides of Danger Zone.* With parties on each side of the ambush confusion may arise as to who shall put in the attack on the enemy, and precious time wasted in getting the attack under way. If both parties attack at the same time and without co-ordination a clash between our troops may result. It is suggested, therefore, that the party not yet entered into the ambush is responsible under command of the convoy commander, or if he is not there, the senior vehicle commander present. The party clear ahead of the ambush will stop, marshall, return to the danger zone and exploit the situation as the convoy commander thinks fit. If the convoy commander is not present, then as the senior vehicle commander thinks fit.

42. Inclusion in a convoy of an armoured vehicle has two effects on ambush action:

a. It is able to give good covering fire to the flanking attack.

b. It affords protection to those caught in the danger zone. Usually the armoured vehicle drives right up to the danger zone and engages the enemy at very short range. See *"The Division in Battle, Pamphlet No 11, Counter Revolutionary Warfare"*

43. The rocket launcher can be concentrated on the enemy position or on their probable escape routes. In order that at least one of these weapons can be operated it is suggested that they should be well spaced out in the convoy.

44. It is possible that the convoy commander may be pinned in the danger zone or find himself with the party on the wrong side of it. In order to ensure there is a nominated commander on the spot to direct an attack, all convoy orders should state that the senior vehicle commander present will organize the assault in the absence of the convoy commander.

45. The above techniques should be practised continually in varying situations until the natural reaction to enemy ambush is the application of a contact drill.

Road Blocks

46. The enemy will sometimes employ road blocks. These will nearly always be covered by troops so that when a vehicle is halted by the block, fire can be quickly directed at personnel riding in the vehicle.

47. Road blocks are usually sited on the brow of a hill or on road bends in such a position that they cannot be detected until the vehicle is on top of them. They take the form of fallen trees, boulders, wire obstacles or land mines. In case the road block has been mined, or booby traps laid in or around it, care must be taken in removing the obstacles.

48. When the leading vehicle stops suddenly there is a tendency for the vehicles in convoy to *"close up"*, thereby increasing the possibility of all vehicles moving into the danger zone before the enemy open fire.

49. All convoys must have a drill for halting immediately a road block is encountered.

50. It is inevitable that there will be many false alarm road blocks, as trees do fall naturally across roads in tropical countries and after rain landslides and rocks may be found. However, all obstacles must be regarded with suspicion until they have been cleared.

51. The drill for road blocks is:

a. The leading vehicle stops and personnel debus quickly.

b. If firing has not broken out one man must be sent back to halt the following vehicles and to report the obstruction.

c. Personnel from the leading vehicles will fan out and search the ground around the block.

d. Rocket launchers will take positions by the side of the road from where they can support an attack.

e. If an armoured vehicle is present, it should move forward and cover the road block from a close fire position.

f. When the ground has been cleared the road block may be removed.

52. If the leading vehicle is attacked before personnel are able to debus then the action will be the same as for a normal vehicle ambush.

Debussing

53. When a vehicle is forced to stop in an ambush position due either to an obstacle or to injury to the driver, troops must debus instantly. This must be taught and practised as a drill.

54. The enemy will attempt to stop the vehicle in the danger zone by use of mines, other obstacles, and by firing at the tyres or driver. Taking advantage of the initial surprise, he then tries to kill all the troops in the vehicle. To do this he normally places an automatic weapon where it can cover the rear of the vehicle.

55. The question of alertness has already been stressed and vehicle commanders must increase alertness whenever likely ambush positions are being approached.

56. To aid debussing all packs and stores will be stacked in the centre of the vehicle. Troops should hold their weapons at all times and spare machine gun belts should be in pouches, not in boxes.

57. When the vehicle is forced to stop:

a. The vehicle commander will shout "*DEBUS RIGHT*" or "*LEFT*" to indicate the direction in which the troops will muster.

b. Sentries fire at anyone seen and in the direction from which enemy fire is coming to disturb the enemy aim.

c. Troops debus over both sides of the vehicle and dash in the direction indicated. As few troops as possible should attempt to debus over the tail of the vehicle.

d. As soon as the troops are clear of the vehicle sentries debus and join the remainder.

e. The aim must now be to collect the fit men as a formed body for counter action. Wounded troops must be attended to after counter action has been taken.

58. The drill must be practised frequently by vehicle loads, eg, infantry sections. Where miscellaneous loads are made before a journey, two or three practices must be held before the convoy moves off.

Training

59. The appearance of a unit's road convoys can tell much about its state of operational efficiency. The enemy can read and interpret the signs and will tend to look for easier targets than those presented by a well trained unit.

60. The danger of ambush must not be allowed to become a bogy. Troops should be taught that, well prepared, they are more than a match for the enemy under any circumstances and that the contact afforded by a road ambush gives an opportunity to close with the enemy and destroy them.

61. When troops are efficient, alert and well disciplined, there is less likelihood of ambush.

62 to 68. Reserved.

SECTION 11 — COUNTER AMBUSH ACTION FOR WATERCRAFT

General

69. Waterways will be used either:

a. As a means to deploy troops into an operational area, or

b. For the logistic support of troops already deployed.

70. These waterways will vary from wide delta type expanses of tidal water to large deep rivers or narrow fast flowing rivers with rapids and high banks.

71. A number of military and local craft will be available for use on these waterways. Generally various types of military

craft, and the larger types of local craft, will be found in the deltas, upstream in the large rivers and the lower tidal areas of the smaller rivers in South East Asia.

Types of Boats

72. This section deals with those boats which can be practically handled in the upper reaches of watershed rivers, and which will normally be a unit responsibility.

73. The provision and handling of larger military or local craft in the lower reaches of rivers is outside the scope of this section.

74. The boats that can be practically handled in most upper reaches are:

a. The Assault Boat Mark IV.

b. Local craft such as long boats and sampans which have been designed for a particular area.

75. A suggested plan for loading troops in a local type craft is described in Annex F.

76. The loading and operation of assault boats is described in *"Field Engineering and Mine Warfare, Pamphlet No 8" Part 1, Annex E.*

77. *Propulsion.* Normally assault and local craft will be equipped with an outboard engine. It is suggested that MT personnel of units be trained in the maintenance and operation of these engines. The soldiers responsible for the operation and the engine and the physical steering of the boat is called the coxswain. It should not be overlooked, however, that the bowman of the long boat is an essential team-mate of the coxswain. His job is to give warning of the obstructions. There will be occasions when silence is required, or engines are not available, when use of paddle is necessary. Pre-operational training in this is a requirement.

Security

78. *Personnel, Weapons and Equipment.* Particularly during an action boats can be overturned and tip their occupants and equipments into the river, therefore certain preliminary security measures are necessary. These are:

a. *Personnel.*

(1) A reliable life jacket should be provided.

(2) Personnel should remove packs and place them in the boats.

(3) Webbing equipment should be loosened, free of shoulder straps and worn over the life jacket, so that it can be quickly discarded.

(4) The ends of trousers should be pulled out of the boots.

(5) Shirts should be buttoned to the neck, and freed of trousers waistbands.

This produces a temporary flotation.

b. *Weapons.* In the event of the boat overturning it is essential that the weapons are recoverable. They should be secured to the boat but in such a way that they can be fired by individuals whilst sitting in the boat.

c. *Engine.* The engine becomes easily detached when the boat capsizes. It is necessary that it is attached by secure means, other than the clamps, to the boat. A special chain attachment can be obtained for this purpose.

d. *Packs and Webbing.* In the event of a boat overturning it must be accepted that packs will normally be lost. Webbing equipment must be discarded and will probably also be lost.

79. Critical points along rivers and waterways should be guarded. Points offering favourable ambush sites should be cleared of vegetation near the banks. A chemical defoliant has proved successful.

80. Other security measures are:

a. Irregular schedules of movement.

b. Mounting automatic weapons on all boats.

c. Providing troops on each craft travelling independently and on each group of craft travelling in convoys.

d. Providing adequate communications for each craft.

e. Establishing waterways patrols in fast, heavily armed craft.

f. Patrolling the waterways by air.

Counter Boat Ambush Drills

81. The problems are very similar to those encountered in motor vehicle ambushes and requires similar immediate action drills. These are:

a. Preparation and preliminary briefing.

b. Convoy system of tactical spacing.

c. Tactical movement.

d. Immediate fire reaction, ie, smoke, LMG and Energa grenades.

e. Counter action drills by unaffected boats.

f. Mopping up and reorganization.

ANNEX A

THE IMMEDIATE AMBUSH

General

1. There will be occasions when a patrol, without being seen itself, sights an enemy party approaching either on a track, or across a clearing, or in jungle. This is most likely to occur when the patrol has halted and the enemy is on the move. It is obviously better to allow the enemy to approach as close as possible before opening fire on him. If the patrol is on the move there may be time only for a silent signal to be passed through the patrol, to move quietly and quickly into the positions indicated and for the signal to open fire to be given when the enemy reaches the position in which he is most vulnerable.

Principles

2. **a.** Good battle discipline is essential to ensure that no shot is fired in the excitement of the moment when the enemy is first sighted.

b. Silent signals must be used.

c. Movement must be quick, silent and concealed where possible.

d. The commander, who should normally spring the ambush, must control the action.

3. Explanatory diagram. See Figure 6.

Drill

4. This is a drill that can be used only when the enemy is moving towards the patrol:

a. The leading scout passes a silent signal back as he sees the enemy approaching and the patrol leader gives the silent signal for immediate ambush.

b. In some cases (such as when the enemy is very close before the leading scout sees him) there may not be time

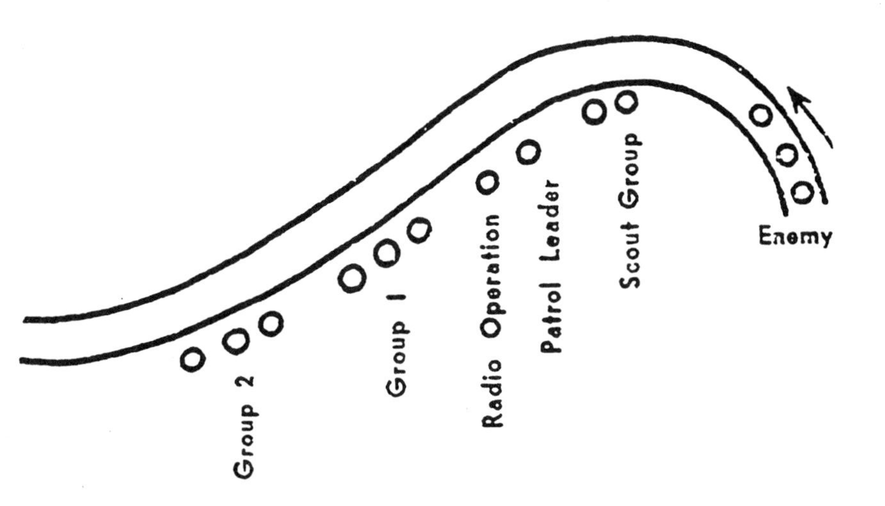

Figure 6 — The Immediate Ambush.

for this. The patrol leader must, therefore, be prepared to delegate to the leading scout the responsibility for giving the silent immediate ambush signal.

c. On seeing the signal, the leading group must get under cover from view and remain still, if they have not assumed a good fire position. The enemy may be too close for their positions to be adjusted.

d. Other groups further down the column will have more time to choose good positions, and the success of the ambush will depend on the machine gun being well sited.

e. The signal to open fire will normally be given by the patrol leader, who should be able to see when the machine gun has a good target. Nevertheless everyone must be ready to open fire if the enemy becomes aware of the ambush before the signal is given. Men will remain in position until ordered to move by the patrol leader.

5. Subsequent action by the patrol leader and ambush party will be similar to the deliberate ambush, if time and opportunity permits.

6. Other formations and circumstances will demand different drills for the immediate ambush.

ANNEX B

AMBUSH ORDERS — AIDE MEMOIRE

(Remember security — Do not use the telephone — Do not allow men out after orders,

Situation

1. a. *Topography*. Use of patrol reports, air photographs, maps and local knowledge. Consider use of a guide.

   b. *Enemy*.

      (1) Expected strength.

      (2) Anticipated order of march.

      (3) Dress and weapons of individuals.

      (4) Habits, including reaction to previous ambushes.

   c. *Friendly Forces*.

      (1) Guides to accompany.

      (2) What other forces are doing.

   d. *Clearance*.

      (1) Challenge.

      (2) Password.

      (3) Identifications.

   e. *Civilians*.

      (1) Locations.

      (2) Habits.

Mission

2. This must be clear in the mind of every man.

Execution

3. a. Type of layout.
   b. Position and direction of fire of groups.
   c. Dispersal point.
   d. Weapons to be carried.
   e. Composition of groups.
   f. Timings and routes.
   g. Formations during move in.
   h. Orders for springing.
   i. Distribution of fire.
   j. Use of grenades.
   k. Action on ambush being discovered.
   l. Orders for immediate follow up if possible.
   m. Orders for search.
   n. Deliberate follow up.
   o. Rendezvous.
   p. Dogs — if any.
   q. Deception plan.
   r. Signals for.
      (1) Enemy are approaching.
      (2) Open fire.
      (3) Cease fire.

(4) Assault.

(5) Searching group out.

(6) Withdraw.

(7) Abandon position.

Administration and Logistics

4. a. Use of transport to area, if any.

b. Equipment and dress including footwear for move in.

c. Rations, if any.

d. Special equipment.

(1) Night lighting equipment.

(2) Explosives.

(3) Cameras. } If insurgent identification required.

(4) Fingerprint equipment. }

e. Medical.

(1) Evacuation plan.

(2) First field dressing, first aid packs.

(3) Medical orderly.

(4) Stretchers.

f. Reliefs.

g. Administrative area if required; orders about cooking, smoking.

h. Transport for return journey.

i. Inspection of personnel and equipment.

(1) Men with colds not to be taken.

(2) Zeroing of weapons correct?

(3) Is ammunition fresh?

(4) Are magazines properly filled?

(5) Insect repellent.

Command and Signal

5. Success signal, if any.

ANNEX C

NOTES ON THE USE OF FLARES

Trip Flare 2/1

1. This piece of equipment is a standard issue, and requires no description. Flares may be used singly or, in improvised methods, in groups such as the *Cluster of Three* and the *Chandler Board* which are described further on.

2. *Characteristics.*

a. A single flare throws a bright light to a distance of approximately 20 metres. The light is inclined to flicker.

b. The illumination lasts for 1¼ minutes.

c. They can be ignited either electrically or by means of the trip wire issued with them. (When improvised methods are used, they will be ignited electrically). Several flares can be ignited using only one tripwire.

d. They are light and easy to carry.

e. They become damp and deteriorate easily.

3. *Employment.*

a. Its main uses are in:

(1) Ambush.

(2) Defence.

b. In ambushes, trip wires are very liable to be sprung by falling branches and animals, and the electrical method of ignition is recommended. Trip flares are extremely useful as silent sentries in dead ground. In this case, ignition by trip wire will be unavoidable.

4. *Method of Ignition.*

a. *Tripwire.* This is described in the instructions which accompany each flare.

b. *Electrically.* A No 33 (electric) detonator is taped to the top of the flare pot, so that the closed end of the detonator is central on it. Connect a pair of leads to the wires on the detonator and run them back to a battery.

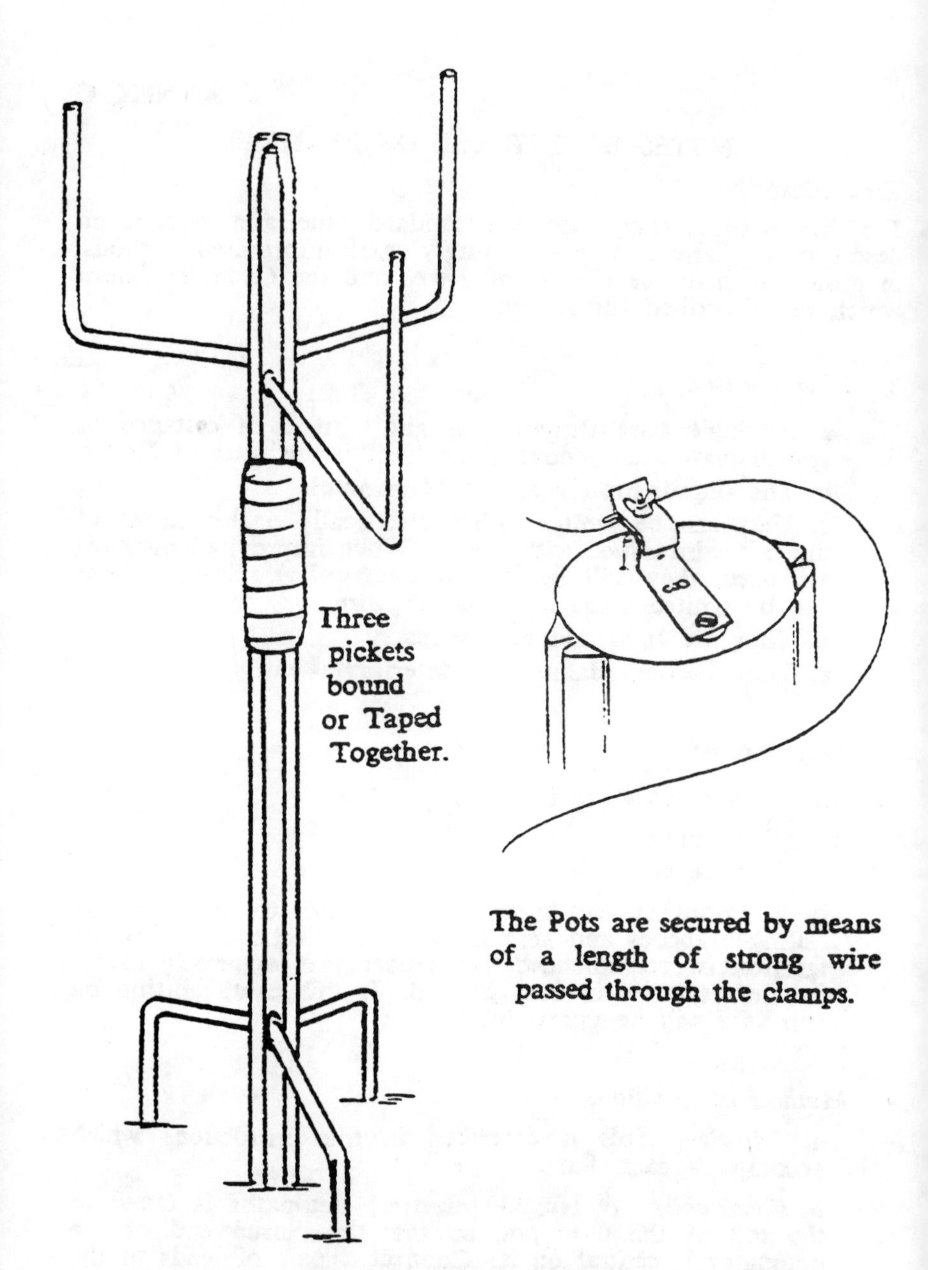

Figure 7 — Night Illumination — Cluster of Three.

5. *Siting.*

a. The flare should be sited so that it is shaded from the eyes of the troops using it, since there is a natural tendency to look at a light source which appears in darkness. This results in a reduction of vision when looking away from the source.

b. The flare should be at waist height from the ground for best effect, which will often mean lashing the spike to a stake which has been driven into the earth.

c. The flare should be camouflaged.

Cluster of Three

6. This is one of the improvised methods of using the trip flare. Three flare pickets are bound together below the brackets which hold the flare pots and the pot is placed on each bracket in the same way as a single pot is placed when one trip flare is used. The three flare pots are tied together by tape, wire, string, etc, bound round the outside so that the whole is secure (See Figure 7). The stakes are then driven into the ground in the required position.

7. *Characteristics.* As for the trip flare, except that the light is thrown to a distance of 60 metres. The best results are obtained when the cluster is at waist height.

8. *Employment.* As for the trip flare.

9. *Method of Ignition. Electrically* A piece of detonating cord approximately 5 ins long is bent into horseshoe shape so that when in position it covers the centres of the lids of the three flare pots. These are not removed or pierced. The cord is then taped so that it is in close contact with the lids, and a No 33 detonator is taped to one end of it. (See Figure 8).

Waterproofing may be effected by using polythene bags from 24 hour ration packs.

10. *Siting.* As in Paragraph 5.

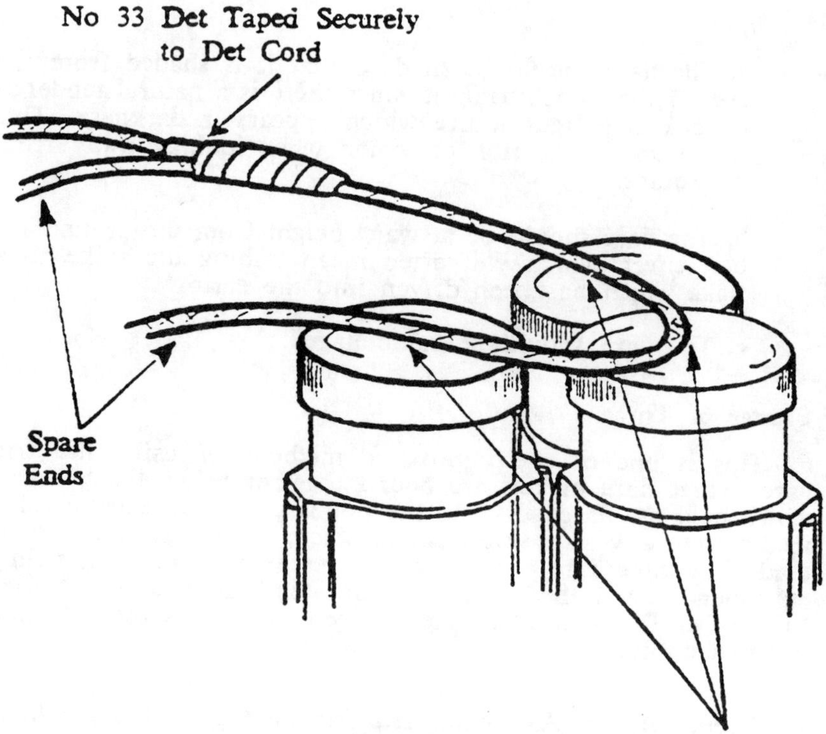

No 33 Det Taped Securely to Det Cord

Spare Ends

Lay Det Cord over the centre of the Flare Cap and tape firmly in posn.

Figure 8—Cluster of Three. Method of Ignition

Chandler Board

11. Another improvised method of using the trip flare, which is basically the same as the Cluster of Three except that the flares are attached to a board, which acts as a reflector. They are attached, so that the bases touch the board, either by binding on with wire, etc, or by assembling as a cluster of three and then driving the stakes through the board.

12. *Characteristics*. As for the Cluster of Three.

13. *Employment*. As for the trip flare.

14. *Method of Ignition.* As for the Cluster of Three.

15. *Siting.* As in Paragraph 5, except that the best results are obtained when the Chandler Board is at head height.

The Verey Pistol

16. The three types of cartridges for the Verey pistol may all be used to give illumination, the illuminating cartridge naturally giving the best result.

17. *Characteristics.*

a. The cartridges are extremely susceptible to damp, making their use very uncertain.

b. The light is poor, and the fast moving shadows which are produced make observation very difficult.

18. *Employment.* Verey cartridges may be used to help in searching ground after the ambush has been sprung. The best results being obtained when the pistol is fired directly overhead.

Hand Flare

19. This piece of equipment is to be adopted shortly. It is hand held equipment which ejects a parachute flare, igniting at a height of approximately 200 metres.

20. *Characteristics.*

a. The burning time of the flare is 35 seconds.

b. It throws a very bright white light which, although steady, causes rapidly moving shadows as the flare falls to earth.

c. It is light and easy to carry.

21. *Employment.* As for the Verey pistol flare.

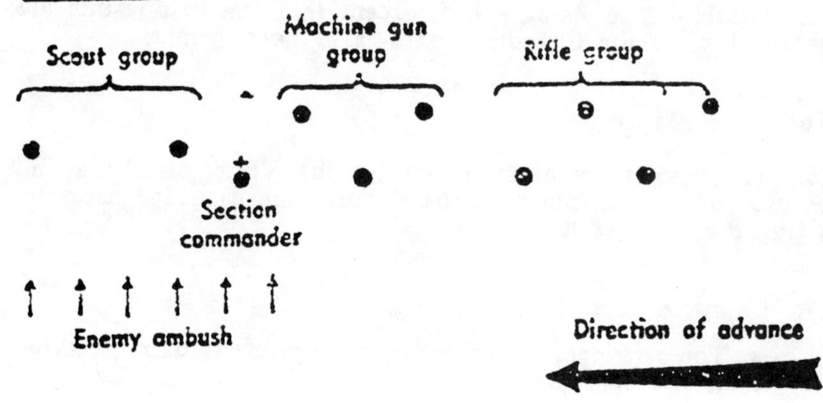
A. On contact
Scout group
Machine gun group
Rifle group
Section commander
Enemy ambush
Direction of advance

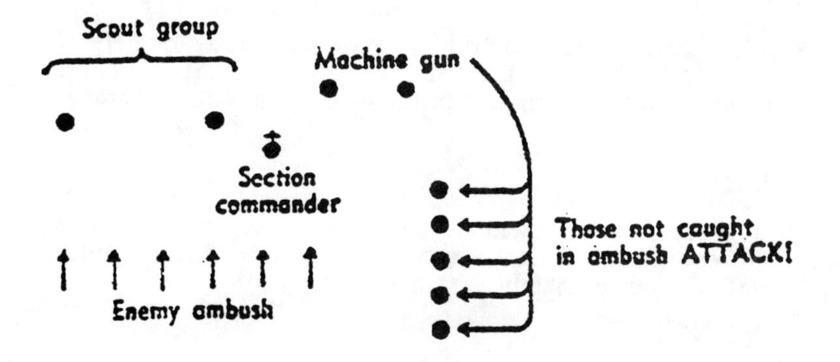
B. After order, "AMBUSH LEFT"
Scout group
Machine gun
Section commander
Those not caught in ambush ATTACK!
Enemy ambush

Figure 9.

ANNEX D

IMMEDIATE ACTION DRILL

1. This drill is required when caught by an enemy ambush. (See Figure 9).

Notes:

> 1. Arrows indicate route taken by members of the patrol who are not pinned to the ground by the opening burst of fire.
>
> 2. Positions of patrol members formed up ready to assault are shown.

2. It helps the speed with which the encircling attack can be put into effect if troops are trained to recognize likely enemy ambush positions and the type of ground he selects for ambushes, and also if they have learnt to recognize the type of fire they will hear if they do run into an ambush.

3. Survivors of the leading group(s) call out "*Ambush Left or Right*" then move to fire positions and engage the enemy. The leading group will always be a tactical bound ahead of the patrol leader.

4. This drill can be worked in reverse if the enemy allows the leading elements of a patrol to pass and takes on its rear elements.

5. The time taken to encircle will vary according to the nature of the ground and the extent of the enemy position but will rarely be less than ten minutes.

6. Commanders select the direction of movement for the assault according to the nature of the ground.

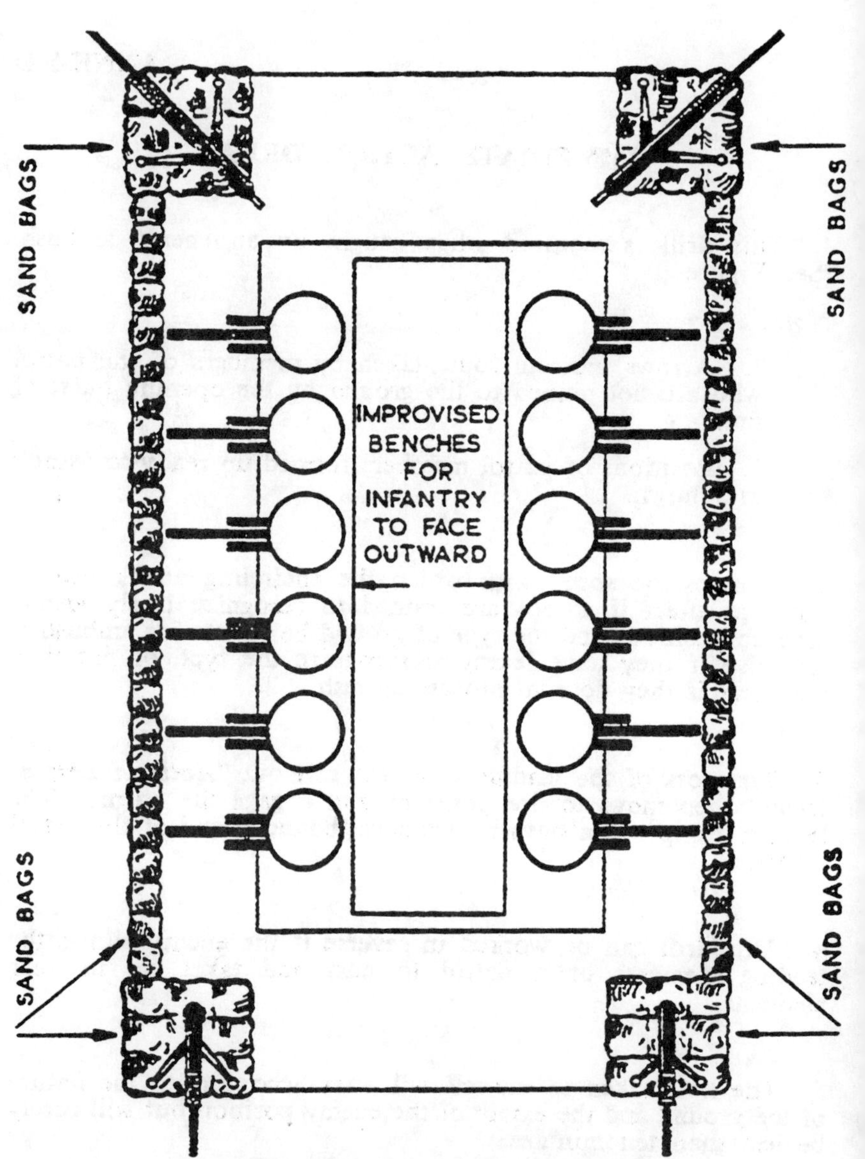

(SAND BAGS ON 4 CORNERS OF TRUCK BED FOR MOUNTING CREW SERVED WEAPONS)

Figure 10.

ANNEX E

NOTES ON THE USE OF SANDBAGS FOR VEHICLE PROTECTION

(See Figure 10.)

Notes:

1. A single row of sandbags, stacked five high is placed down each side of the truck. Firing tests indicate that this single row of sandbags will stop most small arms fire.

2. The troops sit, facing outwards, on a wooden bench set in the centre of the truck bed.

3. Ideally automatic weapons should be positioned at each corner. If these weapons are not available, at least two should be carried — one firmly sandbagged in place on the cabin top and one at the rear.

4. Packs should be hung over the side to offer additional, but limited, protection and concealment.

5. Approximately 70 filled sandbags are required for each truck. The average weight of each bag is about 40 pounds, for a total weight of 2,800 pounds. Troops and equipment weigh an additional 2,300 pounds, thus giving a total vehicle load of 5,100 pounds. This load permits cross country operations without undue wear on the vehicle.

ANNEX F

LOADING OF LOCAL WATERCRAFT

General

1. Local boats vary in length and the corresponding width of beam, according to the types of river in which they are used. Many are very unstable and will capsize if clumsily handled. In some, as many as fifteen fully equipped men can be carried. In others the limit will be four or five.

Loading Factors

2. Before loading the following factors should be remembered:

a. There will nearly always be two men already in the boat, ie, the coxswain or the man who operates the engine and steers the boat, and his mate or bowman who watches for obstructions and assists in navigation.

b. It is easier to load the boat if it is partially beached rather than in deep water.

c. The centre of gravity must be kept as low as possible and stores should be spread along the length of the floor and not piled. Similarly, troops must sit on the floor and not on top of packs, ration boxes, petrol tins, etc.

3. During loading the boat crew must be in attendance to steady the boat. The procedure must be controlled and unhurried.

4. Troops due for embarkation should:

a. Have donned a life jacket.

b. Have their web equipment slung loosely over their shoulders but not fastened in front nor under the shoulder straps.

c. Have their pack in one hand and their weapon in the other.

d. Remove the bottoms of their trouser legs from their boots.

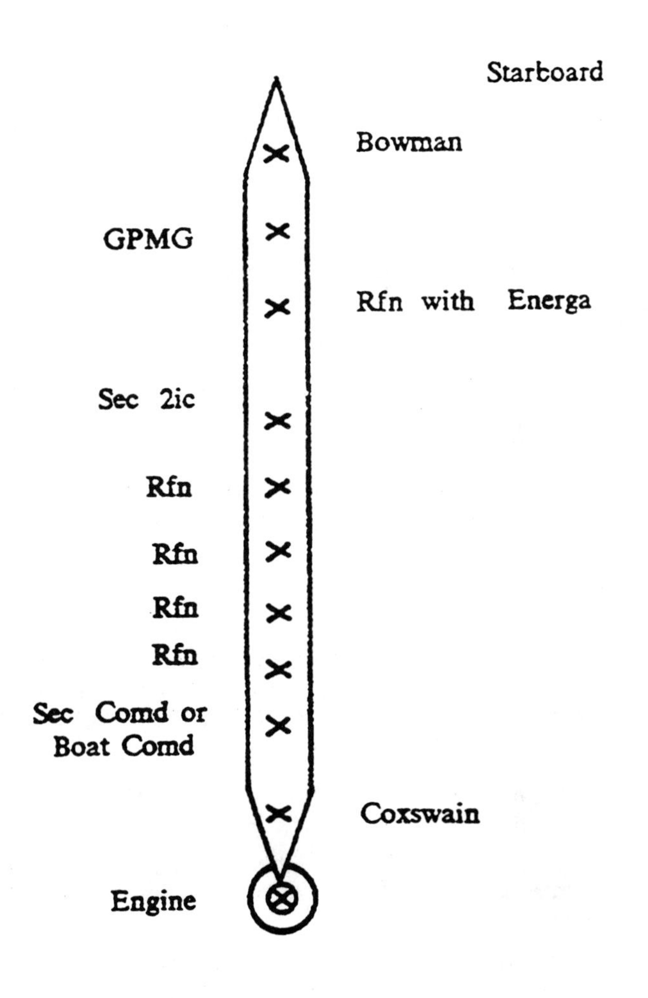
Starboard
Bowman
GPMG
Rfn with Energa
Sec 2ic
Rfn
Rfn
Rfn
Rfn
Sec Comd or
Boat Comd
Coxswain
Engine

Figure 11.

5. The order of loading should be to fill up from the stern first irrespective of whether the boat is side on to the bank, in deep water, or bow on and beached in shallow waters.

6. Regardless of the size of the boat and the number of troops in it, it is suggested that certain personnel sit in the following positions. (See Figure 11).

7. Packs should be placed on the floor between the legs of the owners. Weapons are to be secured to the boat or to the man and a means of flotation, and held at the ready for immediate use.

8. Arcs of observation will be detailed for each man.

Unloading

9. Unloading should again be controlled and unhurried, starting from the bow, with the boat handlers in attendance to steady the boat.

Notes:

1. The Section or Boat Commander is to sit in the stern next to the coxswain. From this position he can best give orders to the coxswain and also see the whole of his section in front of him in the boat.

2. A machine gunner should be in the bow immediately behind the bowman.

3. A rifleman with an Energa grenade prepared for immediate action is to be seated directly behind the machine gunner.

4. The remainder of the section should be seated in their section tactical groups along the length of the boat.